THE WEATHER COACH

Coaching You Through a
Career in T.V. Weather

Marcus Dean Walter

CONTENTS

MARCUSWALTER

The Weather Coach
Coaching You Through a Career in T.V. Weather

CHAPTER 1 - INTRODUCTION

Have you ever wanted to pursue a career in television as a T.V. meteorologist? If so, this is the book for you. With this book, I will coach you through the steps of becoming a T.V. meteorologist and finding a successful career. By reading this book, you will learn:

- What a T.V. meteorologist is
- How to start pursuing a career as a T.V. meteorologist
- How to secure your first T.V. weather job
- And so much more

My name is Marcus Walter and as a T.V. meteorologist, with more than a decade worth of experience, I am excited to be a part of your journey to T.V. weather. I hope you are just as excited as I am! If you are ready to become a T.V. meteorologist, let's begin!

CHAPTER 2 - STARTING YOUR JOURNEY TO T.V. METEOROLOGY

As we begin, I think it is very important to define a T.V. meteorologist, also known as a broadcast meteorologist. We'll start with the base definition of a meteorologist. According to the American Meteorological Society (AMS), a meteorologist is:

-   an individual with specialized education who uses scientific principles to explain, understand, observe or forecast Earth's atmospheric phenomena and/or how the atmosphere affects Earth and life on this planet. There are some cases where an individual has not obtained such a degree but has related educational requirements and has at least several years of professional experience in meteorology

For the purpose of this book, the simplified definition of a meteorologist is someone who forecasts or predicts weather. As well, for the purpose of this book, a T.V. meteorologist is someone who forecasts and presents weather information to the public via television, radio or any other form of media, social media included.

Based on the AMS definition of a meteorologist, in order to officially become a meteorologist, you need to earn a bachelor's degree in meteorology or atmospheric science. While you don't technically need a degree to work in

T.V. or other media, it is encouraged to earn a degree in meteorology so that you can officially call yourself a meteorologist and be prepared to work in this industry. Like with most degrees, you can pursue this degree once you have graduated high school or have earned a GED, General Educational Diploma. Although, you can always take college courses, even meteorology courses, at a local community college, while in high school or earlier.

One of the first things I recommend doing if you are interested in becoming a T.V. meteorologist is to find and read a general weather book. You can find one online, at a bookstore, or borrow one from the library. Reading a book about weather can teach you about the science of meteorology and let you know if you are truly interested in the subject. From here, I would encourage you to find other books and online resources to continue learning about weather. I would then reach out to professionals working in the field on websites such as LinkedIn, X, Facebook, and others, to ask them about their experience working as a meteorologist. All of this should help with figuring out if T.V. weather is for you.

As well, I think it important to talk about T.V. stations and T.V. markets at this juncture. Typically, T.V. stations have a number of staffers working to produce newscasts that reach viewers. These staffers include the news director and assistant news director, who run the newsroom operations, producers who work directly under the news directors, producing, editing video and writing scripts for the newscasts, and reporters and anchors, who report and present the news to the public during newscasts. You have sports anchors and reporters, who cover and present sports news during newscasts. You also have meteorologists, who forecast and present weather information to the public during newscasts.

Within the weather department of the station, you typically have 3 to 4 meteorologists, one of them being the chief meteorologist, who is responsible for leading weather coverage for the station, keeping staffers up-to-date on the latest impactful weather and mentoring and developing other meteorologists on the team. At a smaller market station, you may only have three meteorologists on staff. At a larger market station, you may have 6 or 7 meteorologists on staff. The amount of meteorologists working at a station can also depend on how many newscasts the station produces, i.e. the more newscasts, the more staff needed for those newscasts. For a four person weather team, the chief meteorologist works Monday through Friday, covering the evening shifts. There is a meteorologist that covers the morning shifts from Monday through Friday. There is a meteorologist that covers the weekend morning shifts, that also covers the midday shifts, Monday through Wednesday. And then there is a meteorologist that covers weekend evenings and also midday shifts Thursday and Friday during the week. Typically, the chief meteorologist makes the most money on the weather team, followed by the morning meteorologist and then weekend meteorologists. Salaries can vary across the industry, from $30,000 to $40,000, when you are just starting out, to potentially millions if you make it to national news. Compensation is typically based on experience, education and certifications.

In a given T.V. market, or designated market area (DMA),

there are usually 3 to 4 TV stations, although in smaller markets, there are fewer stations, sometimes called duopoly stations, because they manage two or more T.V. networks under one roof. In large markets, there may be more stations. Across the United States, there are 210 DMAs. The largest DMA is the New York City region, with a T.V. household population of more than 7.3 million. The smallest DMA is the Glendive, Montana region, with a T.V. household population less than 4000.

CHAPTER 3 - WHAT TO STUDY IN HIGH SCHOOL AND IN COLLEGE

If you are in high school thinking about a career as a T.V. meteorologist, I would encourage you to take as many science and math courses as you can. Doing this will prepare you for the rigors of pursuing a bachelor's degree in meteorology and for work in the field as a T.V. meteorologist or meteorologist, in general.

The recommended math and science courses include:

- Algebra
- Geometry
- Calculus
- AP Calculus
- AP Physics
- AP Environmental Science
- AP Chemistry
- Computer Programming
- Statistics
- AP Statistics

In addition to the math and science courses, here's a list of other courses I would encourage taking that would help you become a better presenter and communicator, skills necessary to work in television as a meteorologist:

- Art
- Graphic Design

- Acting/Theatre
- Public Speaking
- Debate

These courses will help prepare you to be a scientist and effective communicator, helping you ultimately to your goal of becoming a T.V. meteorologist.

As far as college is concerned, most schools require students to take a certain amount of general education courses and major courses to complete a degree program. Typically, the first 2 years of your undergraduate education are reserved for general education courses. The last two years are reserved for your major courses. For general education courses, I would encourage those interested in a career in T.V. meteorology to take courses similar to those recommended for high school:

- Art
- Theatre
- Public Speaking
- Writing
- Graphic Design

For the required major courses, including the math and science courses, I would encourage students to make sure there courses satisfy the following:

- Atmospheric Dynamics (at least 3 semester hours)
- Atmospheric Thermodynamics (at least 3 semester hours)
- Atmospheric Physics or Physical Meteorology (at least 3 semester hours)
- Synoptic Meteorology (at least 3 semester hours)

- Mesoscale Meteorology (at least 3 semester hours)
- Atmospheric Measurements & Instrumentation or Remote Sensing (i.e. Satellite/Radar Meteorology) (at least 3 semester hours)
- At least 3 semester hours in applied/ specialty meteorology such as: advanced dynamics, agricultural meteorology, air pollution meteorology, applied climatology, aviation meteorology, broadcast meteorology, hydrology or hydrometeorology, physical oceanography, tropical meteorology, and weather forecasting
- Up to 3 semester hours of a synthesizing experience such as work experience, internship, capstone course or research project
- A sequence of calculus (typically 3 courses) that includes: differential and integral calculus, vector and multivariable calculus
- Probability and Applied Statistics
- Physics (a calculus based course with a lab covering fundamentals of mechanics and thermodynamics)
- An appropriate level of coursework or demonstrated competency in computer science in data analysis, modeling, and visualization to allow inferences about the atmosphere; software development; and application of numerical and statistical methods to atmospheric science problems.
- An appropriate level of coursework or demonstrated competency in communication to effectively communicate and interact with scientific, technical, and lay audiences using scientific evidence; discuss and interpret current weather and climate events and forecasts through multiple modalities, including social media; and craft a scientific presentation and

write a scientific report.

- A course covering Earth's climate system (such as a course on climate change or Earth system science)

The meteorology, science and math courses mentioned above should already be incorporated into the schedule of courses required to earn a degree in meteorology as these courses are required by the United States government to work as a meteorologist for the National Oceanic and Atmospheric Administration and the National Weather Service, which is the bar most meteorology program work to meet.

Some of the most challenging courses will be the math and physics courses. These are the courses that typically weed people out of meteorology programs. Some people have found success with these courses by seeking help from tutors or taking these courses at community college colleges. Whatever way you choose, you can get through these courses. In my experience, most meteorology courses are manageable, so I have high confidence you will be able to make it through these courses.

Along with course work, it would be helpful to participate in clubs and organizations that teach you more about weather and broadcasting. I know it's common for students to participate in T.V. production while in high school, and if you could do the same in college, that would be great preparation for work in television. If you can participate in a weather forecasting club or weather-centric group, that would be beneficial as well, as you may be in a position to learn and grow your weather knowledge and forecasting skills and experience. While in college, I participated in my school's campus weather service, learning and forecasting weather for the state of Pennsylvania. I also provided video weather forecasts for a local cable news channel, where I was able to hone my broadcasting skills.

While T.V. internships are not typically available to high school students, they can be available to college students. Most T.V. stations offer internships and some even pay now. You can search for internship opportunities through visit T.V. stations website, as well as doing a search on Google, Indeed and LinkedIn, to name a few. Your main job during your internship is to learn as much as you can, do as much as you can, and potentially get a job offer or be ready to work as a T.V. meteorologist.

Most students pursue T.V. internships starting between their sophomore and junior years, but you can pursue them earlier. Interning at a T.V. station, working with the weather team, will give you first hand knowledge of what it's like to work in the field. It can also give you perspective on what your career may be like as a T.V. meteorologist. Interning a station should also give you an opportunity to record a weather reel, which you will use later to secure your first T.V. job.

For me, I needed to make money during the summers while in college, so I participated in summer research opportunities initially. It wasn't until my last summer doing research that I also interned with a T.V. station. Some of you may be in a similar situation, so working and interning are perfectly normal things to do.

CHAPTER 4 - APPLYING TO T.V. WEATHER JOBS

During the last 3 months of your undergraduate education is when you should start to apply to T.V. meteorology jobs. This is due to the fact that, for most entry level T.V. weather jobs, stations are looking to hire and have a person start sooner rather than later. It's not like other industries where you may apply a year in advance or even get hired a year in advance.

In order to apply for a T.V. meteorologist job, you will need:

- A resume
- A resume reel available to view online
- Customizable cover letter
- 2 to 3 professional references, either from internships, campus jobs or professors

You can find T.V. meteorologist jobs by searching openings on company websites, through Indeed, Google and T.V. specific job websites such as www.tvjobs.com. You can also network with other T.V. meteorologists to learn about openings. As well, you can upload your reel to a platform like Youtube, where you can then submit a Youtube link for your reel with your application.

Once you have applied, you can send a message to the station's news directors or hiring managers to let them know you have applied and are interested in working at the station. You can also reach out to the weather team members at the various stations where you have applied to let them know of your interest in joining the station. I would refrain, if possible, from calling a station, as most stations typically ask in job announcements to not call. They prefer to reach out to you if they are interested.

Then from here, it's a waiting game. You wait to see if a station or stations are interested in you. Sometimes you wait a week, sometimes you wait even longer, it really depends. Don't get discouraged during this time, just keep applying to job openings until you hear back from a station. Sometimes, you may have to apply to 10 jobs or more before you hear anything.

If you get a response from a hiring manager, expect to do either a phone or Zoom interview soon after. During this interview, you would likely speak with the news director and possibly the chief meteorologist as well. Here are some things to do during the interview:

- Be yourself
- Be confident
- Express enthusiasm about working at the station and with the weather team
- Answer all their questions to the best of your ability
- Have a few questions to ask

If all goes well, you may be invited for a station visit. This happens more regularly for larger market stations. During this visit, you will see their facilities, from the newsroom to the T.V. studio and weather center, along with other areas. You most likely will be taken out for a meal with the news directors and possibly members of the weather team. You may even have to audition on the green screen and at the news desk in the studio. If all goes well from here, you would shortly find out if you have an offer.

Sometimes, you may receive an offer from the first phone or zoom interview. I have seen this happen with smaller stations.

Once you get a job offer, it is customary to respond to the offer within two days, unless you ask for more time to decide. I would encourage you to not take too much time in making a decision. This is also your time to do some negotiating if you can.

In negotiations, you can typically ask for:

- More compensation
- More vacation time
- Financial support to attend weather conferences
- Clothing and make-up allowance
- A buy-out clause
- A set schedule
- Seals and Certification Funds

And you can do this negotiation yourself or hire a lawyer to review your contract and negotiate on your behalf. You can also reach out to an agent for help with this, but they usually work with people with at least one contract under their belt, especially if you are in a smaller T.V. market. First contracts are typically 1, 2 or 3 years.

CHAPTER 5 - STARTING YOUR FIRST T.V. JOB

After you have signed your first contract, it's time to get ready for work and a life transition. Stations usually give you about a month to move to your new home and begin work. If a station is short-staffed, they may ask you to start work sooner. As you begin to work in T.V., here are some tips:

- Be a team player
- Know that you don't know everything so be ready to learn
- Be on time, early is best
- Allow yourself time to get better

When you start at a T.V. station, you will likely start with training of all kinds. This would likely include H.R. training, some journalism training, and weather graphics training, to name a few. Take this time to learn as much as you can, especially for the weather graphics, as this will help you through the rest of your T.V. weather career. Take this time to also learn everything you need to be successful on camera presenting weather. As well, make sure you have everything you need to be ready for the first day on camera. This includes clothing, shoes, make-up, and an ifb earpiece for listening. In most cases, the station provides the ifb earpiece for you, but you can also buy a back-up online through retailers like Amazon.

Once you have been given a date to start on-camera, practice until then. You should be able to practice at the station during your training. Also, take time to learn the region you will be covering and how to pronounce the names of things such as streets, cities and counties. This will allow you to effectively communicate to the viewers watching.

Also, familiarize yourself with the type of weather and geography of the T.V. market. This will help with your ability to forecast and learn how the landscape may impact the forecast. Are there mountains, deep valleys, rivers, lakes, anything else that can impact the weather?

And when the day comes for you to start on air, give yourself time to improve and work out the nerves. Most people are nervous when they start work in T.V. Over time, the more you do this type of work, the more relaxed you get.

Your first year will be about learning and improving as much as you can. Additionally, during your first year or first contract, you should meet with your chief meteorologist or news director to set goals for your development. This will help give you direction as you start your career.

Here's a few more tips:

- Stay out of station drama, as with any job
- Have a flexible work schedule, as your schedule may change a lot due to being the new person on the team
- Maintain good professional relationships
- Work to gain as much severe weather experience as

MARCUSWALTER

you can

CHAPTER 6 - SEARCHING FOR YOUR SECOND JOB

This process typically starts about 6 months before the end of your first contract. At this time, you should put together a new resume reel, resume and cover letter and start applying to jobs. You can look for these jobs in the same places you looked before, Google, Indeed, tvjobs.com, networking, etc.

This is also the time where you should initiate a meeting with your station's management about possibly staying at the station. Your news director would be a great person to have this meeting with, as they are likely your main hiring manager. The goal of this meeting would be to express interest in staying at the station and what you hope to gain by staying. This is something that is very personal. Not everyone wants the same thing. Some people may want a schedule change, other people might want a larger raise, and others might want even more. Either way, this is your time to express what you want and to see if the station can meet your wants. Based on what they say, this would let you know whether you should think about staying or moving on.

This is also a good time to think about getting an agent. T.V. agents can be very helpful during the job search and for career planning. Hiring one will cost part of your salary, anywhere from 8 to 12 percent, but it may be worth it if you reach your ultimate goal, whatever that may be. You can still navigate this field without an agent, so agents are not required. Similar to what was mentioned with a first contract, hiring a lawyer to review and negotiate a contract can be helpful as well. This will cost, but likely less so than what you would have paid an agent.

Once you have applied and received a new job offer, and assuming it's better than what you get from your employer currently, you can negotiate with your station to see if they can match or offer something better. Based on the response from the station, this will let you know if you should accept the new offer and leave or stay with your current station. Ultimately, you want to make the best decision for you.

This process of applying to jobs within 6 months of an end of a contract will be repeated for the rest of your career working in T.V. One thing to note, typically, you make more money if you move around. This has been shown in many industries.

CHAPTER 7 - STARTING YOUR SECOND JOB

When you start your second job, many of the same things from when you started your first job apply:

- Learn as much as you can
- Improve as much as you can
- Be a team player
- Avoid drama

In this role, you want to seek leadership opportunities. Is there something you can manage for the station? Is there something you can take the lead on for the weather team? You also want to continue getting severe weather experience. Gaining this experience will make you a better meteorologist and allow you to gain more career opportunities.

At this time you can also start thinking about what type of life you want as a meteorologist.

- Do you want to stay long term at your current station?
- Where do you want to take your career?
- Do you want to live close to family?
- How do you obtain a good work life balance?

During the second contract, this is where I encourage you to get to weather educational conferences. This will allow you to continue learning meteorology and stay on top of the latest things happening in T.V. weather. As well, this will allow you to meet and connect with fellow meteorologists, helping to build a network within the field. This network would be beneficial as you may be able to find professional mentors and even learn about upcoming job opportunities. If you can get to these conferences sooner in your career that would be recommended as well. I went to my first professional weather conferences when I was in undergrad and continue to go to this day.

And once you have two years full-time or 3 years part-time T.V. weather work under your belt, this is where you can begin to pursue becoming a certified or sealed meteorologist.

CHAPTER 8 - BECOMING CERTIFIED AND SEALED

The American Meteorological Society (AMS) and the National Weather Association (NWA) are two professional weather organizations that have developed their respective seal programs to hold meteorologists to a higher standard while working on T.V. and communicating weather. At one point in the history of T.V. weather, weather wasn't taken as seriously and even used as a gimmick. So, ultimately, the goal of these seal programs are to certify or approve that the holder meets specific educational and experience criteria and has passed the rigorous testing in their knowledge and communication of meteorology needed to be an effective broadcast meteorologist.

While the AMS's Certified Broadcast Meteorologist (CBM) Seal and the NWA's Television Seal of Approval are not required to work in T.V., these seals can be beneficial to your career. Having these seals, or being able to obtain them, can help you secure jobs, as some stations only hire people who have these seals or are eligible to obtain them. Having these seals can set you apart from other T.V. meteorologists in the field and in your T.V. market, especially if they don't have seals. Also having these seals encourage meteorologists to continue to learn meteorology and to stay on top of the latest science in the field. This is done through attending weather conferences and self learning. Lastly, having and pursuing these seals, allow meteorologists to raise the bar for the industry.

In order to obtain the AMS Certified Broadcast Meteorologist Seal or the NWA Television Seal of Approval, you must:

- Hold a bachelor's degree in meteorology or an equivalent degree from an accredited institution
- Worked two years full-time or three years part-time as a T.V. meteorologist
- Apply for one or both seals
- Pass a 100 question, multiple choice weather exam for each seal
- Pass a review of your on-camera weather work for each seal

You can learn more about the AMS Certified Broadcast Meteorologist Seal by visiting www.ametsoc.org. You can learn more about the NWA Television Seal of Approval by visiting www.nwas.org.

CHAPTER 9 - THE NEXT STEPS OF YOUR CAREER

Now that you have earned your AMS Certified Broadcast Meteorologist Seal and NWA Seal of approval, it's now time to plan for what's next for your career. And similar to before, once you enter the last 6 months of your contract with your current station, this is your time to start applying to jobs to see what's out there. Can you make more money, can you have a better schedule? Can you become a chief meteorologist? These are all things to think about as you are finishing a contract. This is also your time to reach out to station management to see what they can offer you. Can you be promoted at your current station? Can you get a bigger raise? Is there room for your career to continue growing at your current station?

Similar to before, once you receive offers to other stations, you can mention this to your station. You don't have to divulge details of the offer, but if it's better than what you currently have at your station, ask your station to match it. Depending on whether an offer can be matched or exceeded will let you know if it may be worth it to move on to the next opportunity. It is important to note here, if you are happy at a station, it's perfectly fine to stay in one place. After a while though, some people get tired of moving around, making staying in one place more appealing.

CHAPTER 10 - WHAT IT TAKES TO MAKE IT TO THE BIG LEAGUES

If your ultimate goal is to make it to a large T.V. market, for instance top 10 or better, here are some things you need to do. First, you need to show a mastery of meteorology and communication. This mastery develops as you work as a T.V. meteorologist. For most top market stations, they want talent to have at least 5 years of strong T.V. weather experience. They also typically want T.V. meteorologists to have their seals or be able to obtain them. Making it to a top 10 market also involves networking with the staff and management of these stations, including producers, news directors and general managers, general managers being those who manage the entire station, so they know who you are and think of you for openings that come up. This would likely involve you going to weather and journalism conferences to meet staffers directly. You may also want to consider getting an agent who can network and pitch you to these stations as well, although, as mentioned earlier, an agent is not required. Making it to a top 10 market will likely involve a little luck and timing as well. Doing all this should help you make it to a top 10 market.

To make it to network T.V. news, working for shows such as the Today Show or Good Morning America, you need to be an excellent storyteller and know how to put stories together, similar to what a seasoned reporter does, along with showing a mastery of meteorology and communications. You also need to be able to handle intense deadline pressure along with an ever changing schedule, especially when it comes to covering big, national weather stories.

To be clear though, many T.V. meteorologists have successful careers in smaller and medium-sized markets as well. So, you don't necessarily need to make it to a large market or network news to earn the pay you want, and to have the life and career you want.

CHAPTER 11 - WHEN THINGS DON'T GO AS PLANNED WITH YOUR CAREER

At times, working in television can be very challenging and stressful. Sometimes, constantly changing schedules can be a challenge. Other times, dealing with other staffers can be a challenge. And sometimes, stations increase your workload, which begins to wear you down. In a worse-case scenario, your station can decide to not renew your contract unexpectedly. Your health may become an issue and limit your ability to work. I have experienced some or all of this. Given what was mentioned, there are some things you can do to make sure you find a job opportunity that is fulfilling and rewarding.

If you find yourself overworked and exhausted from constant schedule changes, you can alway reach out to your chief meteorologist or news director for a solution. If they provide solutions that benefit you, hopefully this would make life better. If they don't provide solutions or the solutions they provided still don't work for you, this is your sign to look for a better opportunity.

If you find that you are not in good health, mentally or physically, many companies have employee assistance programs that can be a resource. You may also need to ask for time off until you are in better health. This may be in the form of a medical leave, usually available to you after a year of work, that also protects your employment, or a general leave of absence, that may not provide employment protections. As with deciding on a job, you ultimately have to make the best decision for you.

If you find yourself looking for a new opportunity, for whatever reason, here are some things to keep in mind. First, make sure to have at least 1 to 2 professional references from your current station that can vouch for you professionally. If you don't have this, you can always use references from other places of employment or volunteering. Second, make sure you regularly save your on-air work to be included in a reel. Preferably, keep this content on your personal computer or online storage drives. Third, have a reason for why you are planning to leave or are no longer with a company. Reasons could include:

- The company decided to go in a different direction
- The company made a decision that was out of my control
- I am looking for a better opportunity
- I took time off for health reasons

Also, be as truthful as possible. You don't have to share any private or personal information. At the same time, know that, if an employer reaches out to a former place of employment, that place of employment is typically only allowed to 1) confirm if you worked there and 2) your dates of employment. Fourth, be able to effectively explain what you bring to the table and what sets you apart from others. Fifth, ask for what you want. This is the best way to get what you ultimately want, whether it's a stable schedule, more money, or anything else.

CHAPTER 12 - OFF YOU GO!

Alright, are you ready to become a T.V. meteorologist? You have been given all the tools necessary to earn a degree in meteorology, find a T.V. internship, pursue a T.V. meteorologist position and find success as a T.V. meteorologist.

This is a fun and exciting career field, one where you really can make a difference in people's lives for the better. And while the career can be challenging at times, it is well worth it. It is possible to become a T.V. meteorologist and you can do just that!

Reach out if you have any comments or questions.
Find me on YouTube as well at www.youtube.com/ TheWeatherCoach

ABOUT THE AUTHOR

Marcus Dean Walter

Marcus Dean Walter is a degreed meteorologist with more than a decade of T.V. meteorology experience. He holds a bachelor's degree in meteorology from The Pennsylvania State University and a master's degree in atmospheric science from Cornell University. He has worked  behind-the-scenes and on-camera as a meteorologist at the local and national news levels. He is originally from the suburbs of Chicago and has lived all across the United States while working as a meteorologist. Feel free to reach out to him on social media on Facebook, Instagram and Twitter. You can find him with the handle @marcusdwalter. You can also find more T.V. weather specific content at The Weather Coach Youtube page, www.youtube.com/TheWeatherCoach.